EXTREME HORRORS

BLACKBIRCH PRESS

An imprint of Thomson Gale, a part of The Thomson Corporation

Detroit • New York • San Francisco • San Diego • New Haven, Conn. • Waterville, Maine • London • Munich

For more information, contact
Blackbirch Press
27500 Drake Rd.
Farmington Hills, MI 48331-3535
Or you can visit our Internet site at http://www.gale.com

Photo credits: Cover: top left and right © Photos.com; top center © Power Photo; middle left © PhotoDisc; middle right © Royalty-Free/CORBIS; bottom left and right © Digital Stock; all pages © Discovery Communications, Inc. except for pages 4, 12 Corel Corporation; pages 8, 32 © Photos.com; pages 1, 16, 20 © Digital Stock; pages 24, 27 © PhotoDisc; page 28 © Royalty-Free/CORBIS; page 36 © PhotoSpin; page 40 © Centers for Disease Control/Dr. Mae Melvin; page 41 (large photo) © Centers for Disease Control

LIBRARY OF CONGRESS CATALOGING-IN-PUBLICATION DATA
Horrors / Sherri Devaney, book editor.
    p. cm. — (Planet's most extreme)
Includes bibliographical references and index.
ISBN 1-4103-0385-3 (hardcover : alk. paper) — ISBN 1-4103-0427-2 (pbk. : alk. paper)
    1. Dangerous animals—Juvenile literature. I. Devaney, Sherri. II. Series.

QL100.H675 2005
590—dc22

2004019715

Printed in the United States of America
10 9 8 7 6 5 4 3 2 1

Be afraid. Be very afraid. We're on a search for the scariest creature on the planet. We're counting down the top ten most extreme nightmares of nature, and finding out why they scare us so badly. Discover the real meaning of fear when terror is taken to its Most Extreme.

10

The **Wolf**

Number ten in the countdown has come straight from our darkest nightmares. The wolf is a scary animal. It's a savage hunter that can bring down an animal five times heavier than a human.

Little Red Riding Hood was not the only human to be frightened by the big bad wolf.

It's no wonder that humans have been afraid of the big bad wolf for thousands of years. Once upon a time wolves and humans hunted in the same forests, and it made sense to be afraid of the mighty wolf. But today, things have changed.

Now we live in houses far away from the ferocious animals that terrified our ancestors. To get a fright today, you have to visit places like movies or amusement parks.

So why do we do it? Why do we go looking for horrifying monsters?

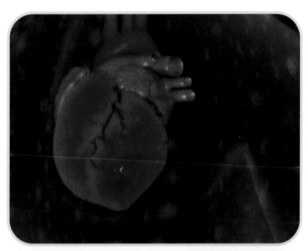

When you're scared, adrenaline floods your body, making your heart race and giving you extra energy to fight or run away.

Being scared is . . . scary. So why do we pay to be terrified by horror movies or roller-coaster rides? The answer lies in the physical nature of fear. To understand why some people are thrill seekers, you have to know what happens to the body when you're scared.

When you're frightened, hormones like adrenaline flood your body. Adrenaline diverts blood from your skin and gut to your muscles. You get extra energy because you begin breathing faster and your heart starts racing. The adrenaline rush even sharpens the senses. Your pupils dilate so you can see better. All these hormonal changes are known as the "fight or flight" response. Thrill seekers love the feeling of their body getting ready to fight for its life—or to run for it.

Although wolves have a reputation as vicious killers, they usually stay away from humans.

The "fight or flight" response wasn't designed for adrenaline junkies. It was meant for a really dangerous situation, like facing a ferocious pack of wolves. Do wolves deserve their terrible reputation? After all, there's never been a documented case of a healthy wild wolf killing a human in North America. In fact, after years of persecution, wolves are scared of people and will keep well out of their way! And that's why the big, bad, bashful wolf is only number ten in our countdown of horror.

The **Mouse**

Would you like a mouse in the house? Incredible as it may seem, the mouse is number nine in our countdown because it's some people's worst nightmare. A recent poll showed 33 percent of American women are afraid of mice! But why? A mouse is harmless, and it's usually so frightened of us that it'll keep well out of our way. But then, some humans can be frightened of the strangest things.

More than 40 percent of Americans are terrified of speaking in public. Some even have a phobia, which means that even though there's no physical danger, their bodies go into the "fight or flight" response. Blood is directed away from their skin, leaving them with cold feet and sweaty hands. Their hearts race and they're ready to fight or, more likely, run screaming from the stage. We sometimes call this phobia "stage fright," but there are many other phobias.

A fear of mice is one of many phobias that people have. One very common phobia is the fear of speaking in public (left).

Chronomentrophobia is the morbid fear of clocks. Triskaidekaphobia is the fear of the number 13, and peladophobia is the fear of bald men!

So why are sufferers of musophobia so afraid of mice? One doctor suggests that it's a fear not of but for the mouse. When a mouse pops up, it stimulates the idea of having to kill it, and that's what causes the anxiety.

Some people have a fear of clocks (top); others are terrified by the number thirteen (middle); while others scream at the sight of a bald man (bottom).

One reason people are so afraid of mice is that these little critters reproduce at a terrifying rate.

The problem is that mice just keep popping up. In good conditions, one female can have up to eleven babies every three weeks. And conditions don't get any better than in the grain belt of Australia. In 1994 a plague of mice in South Australia caused over 65 million dollars worth of destruction. No wonder some Australians are afraid of mice.

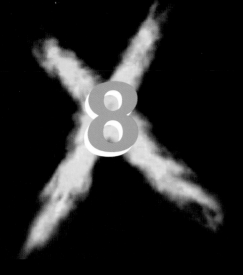

The **Piranha**

Piranhas are little fish with a big reputation. A school of these fish could strip the flesh off a cow in just a few minutes. Piranhas are number eight in the countdown because their ferocity is world famous, even though they're found only in the rivers of South America. You'd certainly think twice about swimming in these waters.

Will this otter be food for a piranha? Actually, piranhas are scavengers, preferring to eat dead fish.

Are piranhas really as horrific as their reputation suggests? Surely any animal that entered the water would be instantly attacked?

Surprisingly, animals such as the otter are safe. It's the dead fish that are in trouble. Contrary to popular opinion, piranhas are scavengers, which is why they're happier eating a dead fish than a live otter. Researchers have found that piranhas are mostly harmless, preferring to feed on the tail fins or scales of fish rather than devouring the critters whole. In fact, in the rivers of South America there's another fish that's far more horrific than the piranha.

Attracted to the scent of urine, the vampirelike candirú can enter inside the body, where it feeds off blood.

Explorers traveling through the Amazon have long heard stories of a tiny catfish called the candirú. Like a tiny vampire it swims up into the gills of a larger fish and feeds on its blood. The truly horrific thing about the candirú is that sometimes it makes itself at home inside humans. It's attracted to urine, so it enters our body through our most private parts!

Never use the Amazon River (above) as your toilet, or you might be inviting a candirú to live inside of you (right)!

If you're traveling in South America, never, ever use the river as a toilet. In South America, it pays to remember that there are all sorts of reasons not to go in the water.

7

The **Bear**

If you visit the woods, you could be in for a big furry surprise. Number seven in the countdown is the bear. One of the biggest is the grizzly bear, the undisputed heavyweight champion of the woods, and one of the biggest carnivores on four legs. It can grow to be 8 feet tall and weigh more than a 1,000 pounds!

One swat from this grizzly's paw can kill. Taking no chances, a conservation officer (inset) carries a tranquilizer gun.

Wildlife conservation officers from the Pennsylvania Game Commission spend their time looking for another, smaller type of bear—the black bear. These armed professionals take serious precautions when walking in the woods. They arm themselves not with bullets, but with darts.

17

A conservation officer examines a bear cub (above). Unlike an adult bear's claws, a cub's are razor sharp (inset).

The officers locate a den where they hope a mother black bear is sleeping peacefully. A tranquilizing injection should ensure that the hibernating mother doesn't wake up. What they want to examine are the mother bear's cubs. Their job is to monitor the bear population, but it's not easy. Even baby bears are armed and dangerous. Officer Scott Tomlinson explains:

The claws of a bear cub are very sharp. In the adults the claws are much larger but not nearly as sharp. The adults wear down their claws by using them for things such as tearing open stumps and turning over rocks. The jaws of a bear are also powerful due to its muscle structure and can easily crush bone.

The ferocious bear is not just a camper's worst nightmare. It's also a kid's best friend. The teddy bear is the best-selling toy of all time, but even the cuddliest teddy bear can be dangerous! Researchers have discovered that nine out of ten teddy bears are contaminated with bacteria. There's no evidence that toy bears are infectious, but there's been a call to ban the teddy bear from doctors' waiting rooms!

An adult bear could kill a human with one swat of its mighty paw, and yet teddy bears are loved more than any other toy animal. That's why bears are only number seven in our extreme countdown.

This boy could be cuddling dangerous bacteria as he squeezes his teddy bear tightly.

The **Shark**

During the summer of 1975, the beaches of North America were quieter than usual. The water was warm, and the weather was fine. But people were frightened to get in the water. They were scared of number six in our countdown, the shark. A fearsome fish was the star of the movie *Jaws*, and it is still scaring people today in the back lots of Universal Studios.

Real great white sharks are notoriously difficult to work with on a sound stage, so the makers of *Jaws* built an animatronic shark and nicknamed it "Bruce" after Steven Spielberg's lawyer! To find out why Bruce was so scary, you just have to take a close look at a real great white.

Even if you were surrounded by a cage of steel it would be a little unnerving being in the water with two tons of muscle and teeth that can move five times faster than an Olympic swimmer! The shark is a big fish with an even bigger appetite.

Perhaps the movie *Jaws* was so successful because it touched on some deep, almost universal anxiety about sharks. We're all a little paranoid when we get into the water, according to Sally Carson, a marine educator at the Aquarium of New Zealand. She explains:

> *The dorsal fin projecting through the surface is a sight that causes great panic among people. They imagine a big, dangerous shark below the surface with great big teeth and immediately rush to get out of the water. As humans, we can't swim very well under water. We can't breathe under water. We have to put on all this extra gear to allow us to breathe under water. Maybe that's why we are quite scared of sharks.*

Featuring an enormous mechanical shark, the movie *Jaws* plays on our deep-rooted fear of sharks.

The human imagination often turns sharks into monsters. People have been inventing monsters for centuries. Ancient explorers always brought back tales of fantastic monsters from far-off lands. It's part of the human consciousness to invent horrific creatures to explain why these alien worlds were so frightening. The monster still remains the most effective tool for facing our deepest fears.

Bruce, the star of *Jaws*, was a monster, but Bruce was made of plastic and designed to sell movie tickets. Real sharks can be monstrous, but they're not all killers. The biggest shark in the world is the size of a whale, and just as gentle. A whale shark doesn't even have teeth.

And not all sharks are enormous. In fact, out of 350 species, not even half grow any bigger than three feet long. Despite all the shark's bad press, there are probably only 100 shark attacks each year worldwide. In fact, that's why shark attacks make the headlines, because they happen so rarely. Sally Carson shares her views:

> *The ocean is the shark's natural environment. If people are going to enter it, they need to enter it on the shark's terms. The shark is the top predator. Humans are the invader. They need to respect the ocean.*

The shark is mainly a monster of our mind. As we learn more about this fearsome fish, our attitudes are slowly changing. That's why sharks are only number six in the countdown.

The gentle, toothless whale shark is the world's largest shark.

A diver doesn't seem terrified of this tiny shark.

The Bat

The bat comes screaming into the countdown at number five because it has been flying through our nightmares for generations. Even today, despite sophisticated night-vision cameras, bats are still surrounded by a fog of superstition.

Many people are scared of bats because they know so little about them, according to University of Tennessee's professional bat man Gary McCracken:

> Bats are mysterious. Most people don't know very much about bats. They're active at night when people are asleep. They're associated with darkness. They live in caves, under the ground, in abandoned buildings, and in church steeples, places that are seen as spooky. People don't know a lot about bats so they fear them.

Half the things people think they know about bats aren't even true. Most people believe that when a bat flies past your head it's trying to attack you. What people don't know is that when you're walking along, you're stirring up dozens of tiny insects. The bats are attacking the bugs, not you! And they won't get tangled in our hair, no matter how closely they fly past our heads!

Bats have long been part of our nightmares. But so much of what we believe about bats isn't true.

25

Bats are very important to people. These fruit bats pollinate many plants that are beneficial to us.

Luckily for bats, some people are fighting against the superstition and prejudice associated with bats. Susan Kleven is an animal educator in Aubrey, Texas, who is trying to spread the good news about bats. She explains:

> We can overcome our phobia of bats by understanding and learning more about their activities and how important they are to people. Bats that eat insects eat a lot of insects that are harmful to people, and fruit bats pollinate a lot of plants that are beneficial to us.

Fruit bats roost in trees during the day, so they're not very mysterious. They're such a common sight that people don't think they are evil at all. That's why in Asia bats are symbols of good health and good fortune! No wonder the bat is only number five in the countdown.

Not everyone fears bats. A Chinese man stands in front of a door decorated with bats, a symbol of good luck in Asia.

4 The Spider

There's no hiding from number four in the countdown. It can be found in our gardens, our houses, and our sheds. Scaring us when we least expect it is the spider.

Perhaps people are frightened of the way the spider scuttles across the ground on those big hairy legs. Maybe it's the way it stares at them with eight beady eyes, or maybe it's those two horrific fangs. Some psychologists believe that everybody has an instinctive distrust of spiders.

If you have a bad experience with a spider, your built-in paranoia can become arachnophobia. Extreme arachnophobics are compulsive cleaners and will even tape windows shut to make sure that spiders can't get into their homes.

Many people have a fear of spiders. Extreme arachnophobics even store their clothing in spider-proof bags.

With the help of virtual reality, arachnophobic Amber MacNeill confronts her fears as she comes face to face with a virtual spider.

The good news is that it may be possible to cure even the most extreme cases of arachnophobia. At the San Diego Center for Advanced Multimedia Psychotherapy, Brenda Wiederhold has been using some modern technology to help arachnophobic Amber MacNeill. She explains her fears:

> *When I was a little girl I would see a spider and run away. I'd then call my mother if the spider was in the house, and get her to come and get it. Now I live in San Diego 3,000 miles away from my mother, so whenever I see a spider in the house I can't call her to come and get it. I figured it was time to get therapy.*

The first part of the therapy uses technology to help teach clients to relax. Using this technology, MacNeill can see her own "fight or flight" response on the computer screen. Then she can practice controlling her own physiology—even in the face of the enemy! She can even come face to face with a virtual spider!

Safe in the world of virtual reality, Amber MacNeill is able to cope with a computer-generated spider. Eventually, she will test the success of her therapy and confront the real thing.

Learning to overcome her fears through therapy, MacNeill is able to make friends with a real tarantula.

3

The **Bee**

When people plant flowers in their gardens they attract the vicious killer that's number three in the countdown—the bee.

Bees swarm the front of a car. A single bee sting could prove deadly to a person with a severe allergy to bee venom.

You may think the bee is our friend, happily pollinating flowers and giving us honey. But bees kill ten times more people each year than the fiercest grizzlies, gators, and spiders combined! That's because up to 2 million Americans are severely allergic to bee venom. One sting could kill them in less than five minutes!

One bee is bad enough, but when it calls for reinforcements, that's the time to get scared. Bees are number three in the countdown because we're desperately frightened of bee swarms.

Many people are terrified of Africanized bees because these bees attack in far greater numbers than do European bees.

In fact the swarm is here. The legendary "killer bees" swept into the United States in 1990. And with them came hysteria and fear. After all, killer bees are fun in the movies, but how would you like to find a huge swarm hanging in the trees of your backyard? This happened in Phoenix, Arizona. Derek Taylor, from the Africanized Bee Removal Specialists, explains:

> *The reason the Africanized bees are so dangerous is that, when they attack you, they attack you in far greater numbers than European bees. They actually have no more venom in each sting*

than a European bee has, but when they sting you they sting in far greater numbers and for a far greater radius. Also, when you upset a European queen she will send 20 to 30 attack bees no matter how big the colony. The Africanized queen will send about half the colony. That means with these mammoth colonies we can get 10, 20, even 30,000 bees in the air at one time stinging people, and that can be very dangerous.

That's why bees are number three in the countdown. The thought of thousands of animals trying to cause pain can be terrifying.

For some people, taking on a swarm of killer bees would be the ultimate nightmare. But, ironically, people have only themselves to blame. Taylor explains how Africanized bees arrived in North America:

In Brazil years ago people had Africanized bees they were working with to make more honey, but seventeen queens got loose. People said the bees would never make it above the Panama Canal because they could never live through the cold winter in the mountains of Mexico. Africanized bees never store enough honey. Well, the only trait the Africanized bees picked up from European bees was to store more honey, and that's why we have these Africanized colonies of 50,000 to 60,000 bees that are now such a threat.

Not even killer bees can compare to the fear and loathing we have for another deadly animal. For there's nothing sweet about our next contender!

The Snake

2

What scares Americans the most? It isn't the dark, or thunder and lightning, or even flying in an airplane. More than half the population admits to being afraid of snakes, coming in at number two in the countdown. Each year around the world, these deadly reptiles kill an estimated 30,000 people!

Snakes petrify many people. This king cobra is an extremely poisonous snake.

That's why snakes are number two in our countdown of horrors. Even an expert like Peter Taylor from the St. Louis Zoo has a healthy respect for the animal he knows and loves. An adult king cobra has enough venom in a single bite to kill an elephant, so he's extra cautious even when dealing with a young snake.

If an expert snake handler like Taylor is so careful around snakes, it's no wonder the rest of us are petrified by them. Taylor explains why he thinks people are so afraid of snakes:

> If it isn't instinctive, one reason I think people are afraid of snakes is because snakes are so different from ourselves. They are cold blooded, they lack legs, they're low to the ground, stealthy, they don't blink, their tongues are forked—so they're quite something to get used to.

The rattlesnake scares off intruders by shaking its rattle (inset), making a sound that strikes fear into most humans.

One sound you never get used to is the rattlesnake's rattle. It's terrifying. A rattlesnake uses its rattle to scare away intruders. Then it doesn't have to waste venom on animals that are too big to swallow.

Being afraid of snakes makes sense. After all, staying clear of a killer markedly increases your life expectancy. But are snakes as bad as we like to think? Our imaginations turn every slithering thing into a venomous killer, but the truth is that only about 10 percent of snakes in the world are poisonous. But most people can't tell the difference. That's why we're scared of them all. After all, when it comes to snakes, it's always better to be safe than sorry.

Because few people can identify poisonous snakes, even this harmless garter snake makes many people fear for their lives.

1

The **Parasitic Worm**

Creeping into number one in the countdown are worms. But not the harmless garden worms. These are man-eating monsters. Don't think this is just some B-movie plotline. There really are horrific worms that hunt us down.

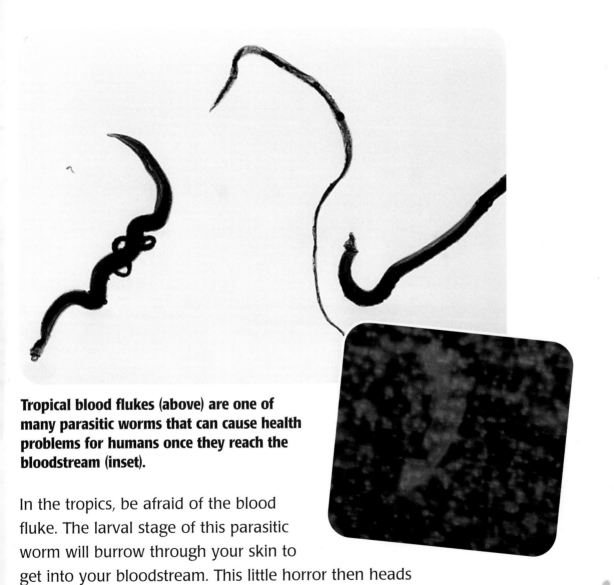

Tropical blood flukes (above) are one of many parasitic worms that can cause health problems for humans once they reach the bloodstream (inset).

In the tropics, be afraid of the blood fluke. The larval stage of this parasitic worm will burrow through your skin to get into your bloodstream. This little horror then heads to the blood vessels near your bladder. The adults can live up to 30 years, constantly laying eggs that are so spiny they can rupture your bladder wall on their way to the outside world!

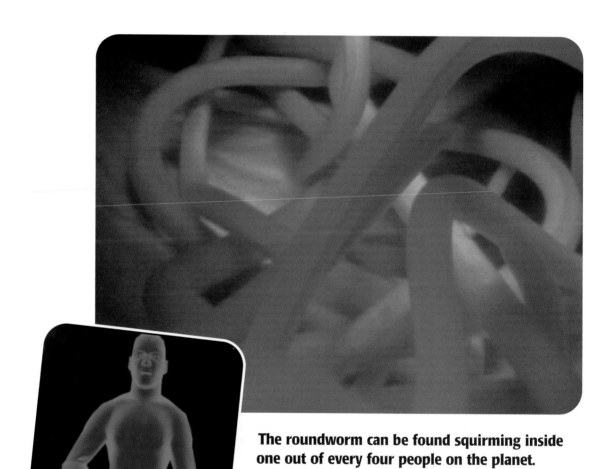

The roundworm can be found squirming inside one out of every four people on the planet.

Our nightmares may be full of huge, ferocious beasts, but the scariest things live inside our bodies! Some live inside our gut—just waiting to feast on the food we send down to them every day. The roundworm Ascaris is as thick as a pencil and twice as long. It's found inside one out of every four people on the planet!

And the scariest thing is that you might never know that these worms are inside you! Usually you only find out when the worm gets upset and decides it's time to leave your body.

Other worms cause trouble because they get stuck in the body. These filarial worms are as thin as sewing thread and as long as a finger. They live coiled up inside lymph glands, feeding on body fluids.

Like other parasitic worms, filarial worms can live in your body without your knowledge.

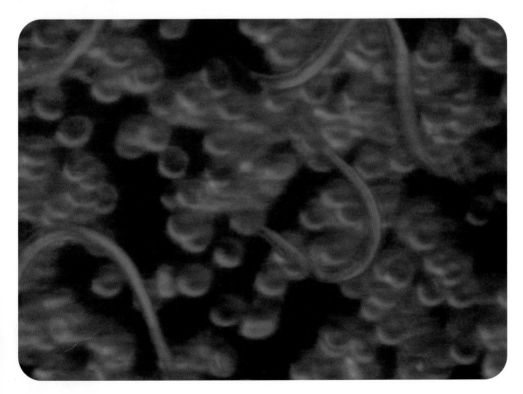

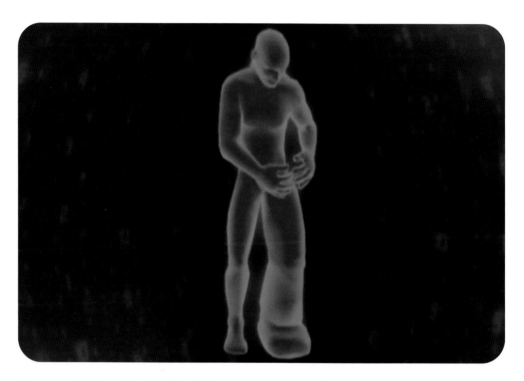

A person infected with filarial worms can experience grossly swollen arms and legs.

The trouble is, all these worms curled up can block fluid movement. And that's how people can get the gross swellings known as ele-phantiasis. How would you like a limb the same size, shape, and color as an elephant's leg?

There's nowhere on your body that worms can't reach. There's even a worm that's at home in your heart! Heartworms are usually para-sites of your pet dog, but sometimes they'll move and live inside your heart, feeding on your blood!

Be afraid. Be very afraid of what may be lurking inside your body. You can't hide from worms. They're everywhere. That's why when it comes to terror, parasitic worms really are The Most Extreme.

Heartworms (below) can live where you might not expect: inside your heart!

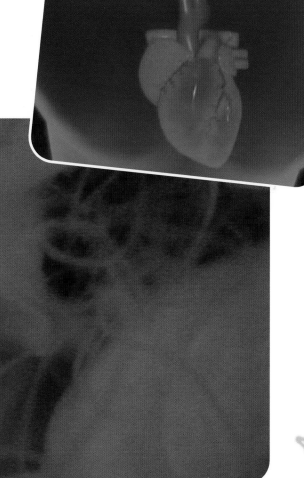

For More Information

Hans D. Dossenback, *Beware We Are Poisonous!: How Animals Defend Themselves*. San Diego: Blackbirch Press, 1998.

Margery Facklam, *What's the Buzz? The Secret Life of Bees*. Austin, TX: Raintree Steck-Vaughn, 2001.

Marcia S. Gresko, *Wolves*. San Diego: KidHaven Press, 2003.

Kris Hirschmann, *Sharks*. San Diego: KidHaven Press, 2002.

Kris Hirchmann, *Snakes.* San Diego: KidHaven Press, 2003.

Elaine Landau, *Creepy Spiders.* Berkeley Heights, NJ: Enslow, 2003.

Elaine Landau, *Killer Bees*. Berkeley Heights, NJ: Enslow, 2003.

Elaine Landau, *Scary Sharks*. Berkeley Heights, NJ: Enslow, 2003.

Patrick Merrick, *Vampire Bats*. Chanhassen, MN: Child's World, 2001.

Darlyne Murawski, *Spiders and Their Webs*. Washington, DC: National Geographic Society, 2004.

Elaine Pascoe, ed., *Snake-Tacular! (The Jeff Corwin Experience)*. San Diego: Blackbirch Press, 2004.

Susan Ring, *Anacondas*. Chicago: Raintree, 2004.

Gloria G. Schlaepfer and Mary Lou Samuelson, *Pythons and Boas.* New York: Franklin Watts, 2002.

Jason and Judy Stone, *Grizzly Bear*. San Diego: Blackbirch Press, 2001.

Glossary

adrenaline: a hormone that helps the body deal with stress

animatronic: a puppet or similar figure that is made active by means of electromechanical devices

arachnophobia: an extreme fear of spiders

bacteria: tiny living cells that can cause disease or do useful things, like make soil richer

carnivore: an animal that eats the flesh of other animals

elephantiasis: a disease in which the skin becomes enormously thickened and swollen

infectious: spread by infection

parasite: an organism that lives in a host organism and injures it in the process

phobia: a fear of a specific object, activity, or situation that causes a person to go to great lengths to avoid it

prejudice: an opinion that is made before all the facts are known

primal: the time when human life on Earth began

roost: to settle down for a rest or sleep

scavenger: an animal that feeds on dead or decaying matter

superstition: a belief that is based on misinformation and fear

therapy: the treatment of a disease of the body or mind

Index